hands-on maths

Kerry Dalton

Year ①

Published by Keen Kite Books
An imprint of HarperCollins*Publishers* Ltd
The News Building
1 London Bridge Street
London
SE1 9GF

HarperCollins *Publishers*
Macken House, 39/40 Mayor Street Upper,
Dublin 1, D01 C9W8, Ireland

ISBN 9780008266950

First published in 2017

10 9

Series Concept and Commissioning: Shelley Teasdale and Michelle I'Anson
Project Manager: Fiona Watson
Editor: Denise Moulton
Cover Design: Anthony Godber
Text Design and Layout: Contentra Technologies
Production: Natalia Rebow

A CIP record of this book is available from the British Library.

Contents

Year 1 aims and objectives

Hands on Maths Year 1 encourages pupils to enjoy a range of mathematical concepts through a practical and hands-on approach. Using a range of everyday objects and mathematical resources, pupils will explore and represent key mathematical concepts. These concepts are linked directly to the National Curriculum 2014 objectives for Year 1. Each objective will be investigated over the course of a week, using a wide range of hands-on approaches such as cubes, 100 squares, dominoes, toy animals, counters and cards. The mathematical concepts are explored in a variety of contexts to give pupils a richer and deeper learning experience, which enables mastery to be attained.

Year 1 programme and overview of objectives

Topic	Week 1	Week 2	Week 3	Week 4	Week 5	Week 6
Counting	Count to and across 20	Count to and across 50	Count to 100	Count in multiples of twos	Count in multiples of fives	Count in multiples of tens
Representing numbers	Identify and represent numbers using objects	Use the language of: equal to, more than, less than (fewer)	Identify and represent numbers using pictorial representations	Identify and represent numbers using pictorial representations	Identify and represent numbers using number lines	Identify and represent numbers using number lines
Place value	Read and write numbers to 20 in numerals	Read and write numbers to 20 in words	Read and write numbers to 50 in numerals	Count, read and write numbers to 100 in numerals	Given a number, identify one more	Given a number, identify one less
Addition and subtraction	Read, write and interpret mathematical statements involving + and = signs	Read, write and interpret mathematical statements involving − and = signs	Represent and use number bonds and related subtraction facts within 20 (to 10)	Represent and use number bonds and related subtraction facts within 20	Add one-digit numbers to 20	Subtract one-digit numbers to 20
Multiplication and division	Solve one-step problems involving multiplication using concrete objects, pictorial representation and arrays	Solve one-step problems involving division using concrete objects, pictorial representation and arrays	Make connections between arrays, number patterns and counting in twos	Make connections between arrays, number patterns and counting in fives	Make connections between arrays, number patterns and counting in tens	Through grouping and sharing small quantities, pupils begin to understand doubling numbers and quantities
Fractions	Recognise, find and name a half of a shape	Recognise, find and name a half of a set of objects	Recognise, find and name a half of a quantity	Recognise, find and name a quarter of a shape	Recognise, find and name a quarter of a set of objects	Recognise, find and name a quarter of a quantity

Introduction

The *Hands-on maths* series of books aims to develop the use of readily available manipulatives such as toy cars, shells and counters to support understanding in maths. The series supports a concrete–pictorial–abstract approach to help develop pupils' mastery of key National Curriculum objectives.

Each title covers six topic areas from the National Curriculum (counting, representing numbers, understanding place value, the four number operations: addition and subtraction and multiplication and division, and fractions). Each area is covered during a six-week unit, with an easy-to-implement 10-minute activity provided for each day of the week. Photos are included for each activity to support delivery.

Hands-on maths enables a deep interrogation of the curriculum objectives, using a broad range of approaches and resources. It is not intended that schools purchase additional or specialist equipment to deliver the sessions; in fact, it is hoped that pupils will very much help to prepare resources for the different units, using a range of natural, formal and typical maths resources found in most classrooms and schools. This will help pupils to find ways to independently gain a deep understanding and enjoyment of maths.

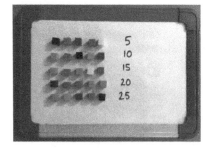

A typical 'hands-on' classroom will have a good range of resources, both formal and informal. These may include counters, playing cards, coins, Dienes, dominoes, small objects such as toy cars and animals, Cuisenaire rods, hundred squares and hoops.

There is no requirement to use *only* the resources seen in the photographs that accompany each activity. Cubes may look like those in the green bowl, or will be just as effective if they look like the ones in the blue bowl. They serve the same purpose in helping pupils understand what the cubes represent.

Resources

Hands-on maths uses a range of formal, informal and 'typical' resources found in most classrooms and schools. To complete the activities in this book, it is expected that teachers will have the following resources readily available:

- whiteboards and pens for individual pupils and pairs
- Dienes and Cuisenaire rods
- dice, coins and bead-strings
- a range of cards, including playing cards, place-value arrow cards and numeral cards

- collections of objects that pupils are interested in and want to count, such as toy cars, toy animals and shells
- bowls/containers to store sets of resources in, making it easy for pupils to handle and use the objects

- ten frames (these could be egg boxes, ice-cube trays, printed frames or something pupils have created themselves)

- number lines and hundred squares – lots of different types and styles: printed, home-made, interactive, digital or practical … whatever you prefer, and whatever is handy. (For hundred squares, there is, of course, the 1–100 or 0–99 choice to make; both work and it is best to choose whatever works for the class. Both offer a slight difference in place value perspective, with 0–99 giving the 'zero as a place holder' emphasis, while the 1–100 version helps pupils to visualise the position of 100 in relation to the two-digit numbers.)

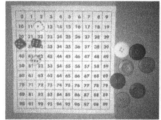

- counters and cubes – lots of them! Many of the activities require counters and cubes to be readily available. The cubes can be any size and any colour: what the cubes represent is the most important factor.

Maths is a truly unique, creative and exciting discipline that can provide pupils with the opportunity to delve deeply into core concepts. Maths is found all around us, every day, in many different forms. It complements the principles of science, technology and engineering.

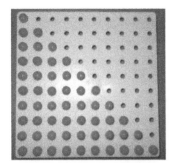

Hands-on maths provides ideas that can be adapted to suit the broad range of needs in our classrooms today. These ideas can be used as a starting point for assessment – before, during or after teaching a particular topic has taken place. The activities are intended to be flexible enough to be used with a whole class and can, of course, be differentiated to suit individual pupils in a class.

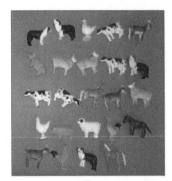

The activities can be adapted to link to other subject areas and interests. For example, a suggestion to use farm animals may link well to a science unit on classification or food chains; alternatively, the resource could be substituted with bugs if minibeasts is an area of interest for pupils. Teachers can be as flexible as they wish with the activities and resources – class teachers know their pupils best.

Spoken language is underpinned in maths by the unique mathematical vocabulary pupils need to be able to use fluently in order to demonstrate their reasoning skills and show mathematical proof. The correct, regular and secure use of mathematical language is key to pupils' understanding; it is the way in which they reason verbally, negotiate conceptual understanding and build secure foundations for a love of mathematics and all that it brings. Each unit in *Hands-on maths* identifies a range of vocabulary that is typical, but by no means limited to, that particular unit. The way the vocabulary is used and incorporated into activities is down to individual style and preference and, as with all of the resources in the book, will be very much dependent on the needs of each individual class. A blank template for creating vocabulary cards is included at the back of this book.

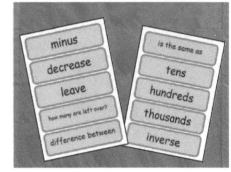

Week 1: Counting

Count to and across 20

Resources: cubes, bowls, objects

Vocabulary: number, numeral, zero, one, two, three … twenty, thirty … one hundred, none, how many?, count, count up / on / down / back, count in ones, twos, fives and tens, many, few, odd, even, every other, how many times?, pattern, pair, ones, tens, digit, 'teen' numbers, first … twentieth, last, before, after, next, between, halfway between

Monday

Give each pair of pupils a bowl of cubes.

Ask them to count the cubes and make a staircase, from 1 cube to 20 cubes. You could place a number line underneath the staircase to demonstrate the numeral linked to the total number of cubes in each tower.

Tuesday

Place a small number of cubes (between 1 and 5) in a bowl.

Ask pupils to count the number of cubes in the bowl; encourage visual counting (subitising). Tell pupils that you are going to count on, out loud, together from that number (e.g. 4) as you drop more cubes into the bowl one-by-one. Add between 5 and 9 cubes so pupils count beyond 10. Ask pupils to say the number you finish counting on.

Wednesday

Place a number of cubes in a bowl (between 9 and 20).

Ask pupils to count the number of cubes and to place a corresponding number of different objects in another bowl.

Thursday

Place a number of cubes in a bowl (between 15 and 25).

Ask pupils to count the number of cubes and to place a corresponding number of different objects in another bowl.

Friday

Provide a selection of interesting objects for counting.

Write a number between 15 and 25 on a whiteboard and ask pupils to count out that number of objects. They should check each other's work in pairs.

You could also roll a dice to enable them to count on from a given number.

Week 2: Counting

Count to and across 50

Resources: 100 squares, objects, bowls

Vocabulary: number, numeral, zero, one, two, three … twenty, thirty … one hundred, none, how many?, count, count up / on / down / back, count in ones, twos, fives and tens, many, few, odd, even, every other, how many times?, pattern, pair, ones, tens, digit, 'teen' numbers, first … twentieth, last, before, after, next, between, halfway between

Monday

Display a large 100 square.

Count in ones, from 1–50, and then count back from 50–1.

Tuesday

Display a large 100 square (or an interactive whiteboard).

Ask a pupil to choose a number between 15 and 25. Point to the number and count together, in ones, from that number to 50 and back again.

Wednesday

Provide a selection of objects in bowls, with between 25 and 30 objects in each bowl.

Working in pairs, ask pupils to count the number of objects in each bowl, removing one object each time as they count.

Thursday

Provide a large selection of objects.

Ask pupils to work in pairs to count out 50 objects. Encourage grouping the objects into tens for ease of counting and checking. Ask pupils to check each other's quantities.

Friday

Give pupils a 100 square.

Call out a number between 25 and 35 and tell pupils that you will count together from that number to 55 (crossing 50) and back to the original number. Practise several times.

Week 3: Counting

Count to 100

Resources: 100 squares, objects, bowls

Vocabulary: number, numeral, zero, one, two, three … twenty, thirty … one hundred, none, how many?, count, count up / on / down / back, count in ones, twos, fives and tens, many, few, odd, even, every other, how many times?, pattern, pair, ones, tens, digit, 'teen' numbers, first … twentieth, last, before, after, next, between, halfway between

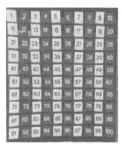

Monday

Display a large 100 square.

Count in ones from 1–100, and then count back from 100–1. Consider splitting pupils into groups so they each count a set of numbers (e.g. one pupil in the group counts from 1–10, the next pupil counts from 11–20, etc.).

You could time pupils and practise this each day.

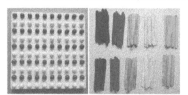

Tuesday

Provide a large selection of objects in bowls.

Pupils work in pairs to count out 100 objects. Photograph the objects to remind pupils what 100 looks like in a variety of forms.

Encourage pupils to group the objects into tens for ease of counting and checking. Ask pupils to check each other's quantities.

Wednesday

Prepare a selection of 'almost 100' objects in bowls.

Tell pupils that you have tried hard to accurately count 100 objects, but that you have made some mistakes and you need their help. Ask pupils to work in pairs to count the objects accurately. Encourage pupils to group the objects into tens for ease of counting and checking.

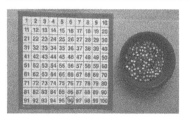

Thursday

Give each pupil a 100 square. Use the objects from Wednesday.

Ask pupils to count the number of objects and then to circle that number on their 100 square.

Friday

Prepare a selection of 'almost 100' objects in bowls.

Ask pupils to count the objects and then compare the quantities to see which is the largest and smallest. Use questioning to ensure understanding of the vocabulary of comparisons (e.g. 'Who has a quantity greater than 95?', 'Does anyone have a quantity equal to 98?').

Count in multiples of twos

Resources: counters, bowls, cups, objects

Vocabulary: number, numeral, zero, one, two, three … twenty, thirty … one hundred, none, how many?, count, count up / on / down / back, count in ones, twos, fives and tens, many, few, odd, even, every other, how many times?, pattern, pair, ones, tens, digit, 'teen' numbers, first … twentieth, last, before, after, next, between, halfway between

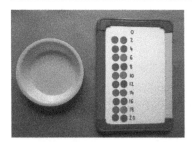

Monday

Give each pair of pupils a bowl of 20 counters (any object would work) and a whiteboard and pen.

Ask pupils to lay out the counters in pairs and write the cumulative total at the side. Highlight the row with 0 counters in.

Count together in twos from 0–20 and back to 0.

Tuesday

Give each pair of pupils 20 counters and 10 plastic cups (or an egg tray, bowls or bun cases, or ask them to draw circles on their whiteboards).

Working in pairs, pupils practise counting out 2 objects at a time and placing them into the cups while counting out loud. Once they reach 20, ask them to remove 2 counters at a time and count back in twos until they get to 0.

They should practise with a range of objects.

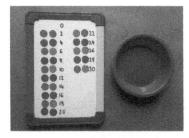

Wednesday

Give each pair of pupils a bowl of 30 counters (any object would work) and a whiteboard and pen.

Ask pupils to lay out the counters in pairs and write the cumulative total at the side. Highlight the row with 0 counters in.

Count together in twos from 0–30 and back to 0.

Thursday

Put up to 30 counters in a number of bowls. Give each pupil a whiteboard and pen.

Ask pupils to count in twos to find the total number of counters in each bowl and to record their count on a whiteboard.

Friday

Give each pair of pupils a bowl of 30 counters.

Working in pairs, pupils practise counting to 30 by placing 2 counters each time into a different bowl.

Then count together as a whole class.

Week 5: Counting

Count in multiples of fives

Resources: objects, containers, cups, 100 square

Vocabulary: number, numeral, zero, one, two, three … twenty, thirty … one hundred, none, how many?, count, count up / on / down / back, count in ones, twos, fives and tens, many, few, odd, even, every other, how many times?, pattern, pair, ones, tens, digit, 'teen' numbers, first … twentieth, last, before, after, next, between, halfway between

---

Monday

Provide a selection of objects for use during the week.

Pairs of pupils count out 30 objects into their container. Ask them to lay out the objects in groups of 5 on a sheet of paper or a whiteboard and to write the numbers underneath. They should circle, highlight or underline the multiples of five.

Photograph the counting pictures they have made to display in the classroom. Keep the containers of 30 objects to use throughout the week.

Tuesday

Give each pair of pupils a container of 30 objects and 6 plastic cups (or an egg tray, bowls or bun cases or ask them to draw circles on their whiteboards).

Ask pupils, in pairs, to practise counting out 5 objects and placing them into the cups. Once pupils reach 30, ask them to remove 5 counters at a time until they get to 0.

They should practise with a range of objects.

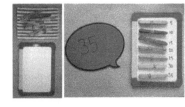

Wednesday

Give each pair of pupils a container of 50 objects, bundled into fives, and a whiteboard and pen.

Call out a multiple of five. Pupils count out that number of objects and then bundle them into fives to count together to check.

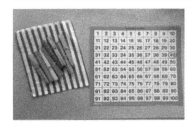

Thursday

Give each pupil a whiteboard and pen. Display a 100 square.

Call out a target number which is a multiple of five and count together in fives using the 100 square as a resource. (Pupils can also write the numbers on a whiteboard if they are confident.)

Friday

Give each pupil a whiteboard and pen. Display a 100 square.

Call out a multiple of five and ask pupils to write out the multiples of five to that number.

Week 6: Counting

Count in multiples of tens

Resources: objects, containers, Dienes sets, 100 square

Vocabulary: number, numeral, zero, one, two, three … twenty, thirty … one hundred, none, how many?, count, count up / on / down / back, count in ones, twos, fives and tens, many, few, odd, even, every other, how many times?, pattern, pair, ones, tens, digit, 'teen' numbers, first … twentieth, last, before, after, next, between, halfway between.

Monday

Provide a selection of everyday objects for use throughout the week.

Working in pairs, pupils count out 50 objects into their container.

Ask them to lay out the objects in rows of 10 and write the multiples of ten at the end of each row. Photograph their counting pictures to display in the classroom.

Tuesday

Give each pair some Dienes tens rods and a whiteboard and pen.

Ask pupils to place the rods on their whiteboard and, starting at 0, to write the cumulative totals from 0–100.

Use the resource to count forwards and backwards in tens.

Wednesday

Give each pair of pupils a bag of Dienes tens rods. Place a selection of multiples of ten everyday objects around the room.

Ask pupils to count the objects into groups of 10, and then place the appropriate number of tens rods next to them. Pupils check each other's work.

Photograph their work to display around the classroom throughout the week.

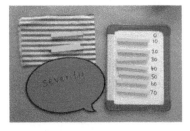

Thursday

Give each pair of pupils a bag of Dienes tens rods and a whiteboard and pen.

Call out a multiple of ten. The first pair to lay out the correct number of tens rods and write the cumulative totals are the winners.

Friday

Give each pupil a whiteboard and pen. Display a 100 square.

Call out a multiple of ten. Ask pupils to write out the multiples of ten to that number.

Week 1: Representing numbers

Identify and represent numbers using objects

Resources: lolly sticks (or similar), numeral cards, coins

Vocabulary: number, numeral, zero, one, two, three … twenty, thirty … one hundred, none, how many?, count, count up / on / down / back, count in ones, twos, fives and tens, many, few, odd, even, every other, how many times?, pattern, pair, ones, tens, digit, exchange, 'teen' numbers, the same number as, as many as, equal to, greater than / less than, less / more, larger, bigger, less, fewer, smaller, compare, order, size, first … twentieth, last, before, after, next

Monday

Give each pair of pupils a set of 50 lolly sticks (or use straws or Dienes).

Using numeral cards, show pupils a number between 1 and 20. Pupils represent the number using lolly sticks that are in bundles of tens and ones. Once they are confident, move on to numbers up to 50.

Tuesday

Give each pair of pupils a set of 100 lolly sticks (or use straws or Dienes; pairs could have different apparatus).

Using numeral cards, show pupils a number between 50 and 100. Pupils represent the number using lolly sticks that are in bundles of tens and ones.

Wednesday

Give each pair of pupils a set of coins (10 × 10p and 10 × 1p).

Write a value between 1p and 10p on a whiteboard. Pupils represent that number using coins.

Next, write a value between 11p and 19p and model exchanging 10 × 1p coins for a 10p coin. Then ask pupils to represent a value between 11p and 19p.

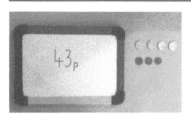

Thursday

Give each pair of pupils a set of coins (10 × 10p and 10 × 1p).

Following on from Wednesday's activity, write a value between 11p and 50p on a whiteboard. Pupils represent that number using coins.

Friday

Give each pair of pupils a set of coins (10 × 10p and 10 × 1p).

Write a value between 21p and 99p on a whiteboard. Pupils represent that number using coins.

Week 2: Representing numbers

Use the language of: equal to, more than, less than (fewer)

Resources: Cuisenaire or number rods, Dienes sets

Vocabulary: number, numeral, zero, one, two, three … twenty, thirty … one hundred, none, how many?, count, count up / on / down / back, count in ones, twos, fives and tens, many, few, odd, even, every other, how many times?, pattern, pair, ones, tens, exchange, digit, 'teen' numbers, the same number as, as many as, equal to, greater than / less than, less / more, larger, bigger, less, fewer, smaller, compare, order, size, first … twentieth, last, before, after, next

Monday

Give each pair of pupils a set of Cuisenaire rods (or number rods) and a whiteboard and pen.

Working in pairs, pupils create as many number sentences as they can using the rods. Pairs should then record using either '9 + 1' or '9 and 1'. Encourage the use of '10 + 0' and '0 + 10' to reinforce place value.

Photograph the combinations for display throughout the week.

Tuesday

Give each pupil a set of Cuisenaire rods (or number rods) and a whiteboard and pen.

Working in pairs, each pupil makes the numbers 10–20, with each partner representing the number in a different way. Ask how many combinations there are. Begin with the use of a 10 + x format so pupils can compare numbers against that model.

Photograph the combinations for display throughout the week.

Wednesday

Give each pair of pupils a set of Dienes and a whiteboard and pen.

Give a target number between 20 and 50. Pupils lay out Dienes to create that number. They must investigate different ways to make the target number. These can be recorded on whiteboards using whole numbers or a combination of tens and ones.

Thursday

Give each pair of pupils a set of Dienes and a whiteboard and pen.

Give a target number between 50 and 100. Pupils use Dienes to represent a number that is **less than** the given number. They must investigate different ways to make the target number. These can be recorded on whiteboards using whole numbers or a combination of tens and ones.

Friday

Give each pair of pupils a set of Dienes and a whiteboard and pen.

Give a target number between 50 and 100. Pupils use Dienes to represent a number that is **more than** the given number. They must investigate different ways to make the target number. These can be recorded on whiteboards using whole numbers or a combination of tens and ones.

Week 3: Representing numbers

Identify and represent numbers using pictorial representations

Resources: place-value arrow cards, objects

Vocabulary: number, numeral, zero, one, two, three … twenty, thirty … one hundred, none, how many?, count, count up / on / down / back, count in ones, twos, fives and tens, many, few, odd, even, every other, how many times?, pattern, pair, ones, tens, exchange, digit, 'teen' numbers, the same number as, as many as, equal to, greater than / less than, less / more, larger, bigger, less, fewer, smaller, compare, order, size, first … twentieth, last, before, after, next

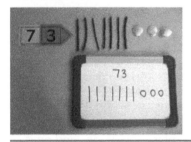

Monday

Give each pair of pupils a selection of natural objects (which they may enjoy collecting) and a whiteboard and pen. They will need different objects to represent tens and ones.

Show pupils a number between 1 and 20 using place-value arrow cards and ask pupils to represent the number using their chosen objects. Pupils record using pictures.

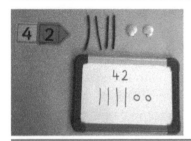

Tuesday

Give each pair of pupils a selection of natural objects and a whiteboard and pen. They will need different objects to represent tens and ones.

Show pupils a number between 20 and 50 using place-value arrow cards and ask pupils to represent the number using their chosen objects. Pupils record using pictures.

Wednesday

Give each pair of pupils a selection of natural objects and a whiteboard and pen. They will need different objects to represent tens and ones.

Show a number between 50 and 100 using place-value arrow cards and ask pupils to represent the number using their chosen objects. Pupils record using pictures.

Thursday

Invite each pair of pupils to choose two different sets of objects to represent tens and ones (to allow them to repeat Tuesday's activity using different objects). Give each pair a whiteboard and pen.

Show pupils a number between 20 and 50 using place-value arrow cards and ask pupils to represent the number using their chosen objects. Pupils record using pictures.

Friday

Invite each pair of pupils to choose two different sets of objects to represent tens and ones (to allow them to repeat Wednesday's activity using different objects) and a whiteboard and pen.

Show a number between 50 and 100 using place-value arrow cards and ask pupils to represent the number using their chosen objects. Pupils record using pictures.

Week 4: Representing numbers

Identify and represent numbers using pictorial representations

Resources: 100 square, objects

> **Vocabulary:** number, numeral, zero, one, two, three … twenty, thirty … one hundred, none, how many?, count, count up / on / down / back, count in ones, twos, fives and tens, many, few, odd, even, every other, how many times?, pattern, pair, ones, tens, exchange, digit, 'teen' numbers, the same number as, as many as, equal to, greater than / less than, less / more, larger, bigger, less, fewer, smaller, compare, order, size, first … twentieth, last, before, after, next

Monday

Make available a range of objects for pupils to choose from. Give each pupil a whiteboard and pen.

Show pupils a number on a 100 square. Ask pupils to show a representation equivalent to this number using objects of their choosing. They draw the objects. Repeat to practise. Photograph for the next day as a reminder.

Tuesday

Remind pupils of Monday's activity using the photographs. Give each pupil a whiteboard and pen.

Show pupils a number between 20 and 50 on a 100 square. Ask pupils to show a representation of a number **less than** the number given, represented by a picture. Repeat to practise. Photograph for the next day as a reminder.

Wednesday

Remind pupils of Tuesday's activity using the photographs. Give each pupil a whiteboard and pen.

Show pupils a number between 50 and 100 on a 100 square. Ask pupils to show a representation **less than** the number given, represented by a picture. Repeat to practise. Photograph for the next day as a reminder.

Thursday

Remind pupils of Wednesday's activity using the photographs. Give each pupil a whiteboard and pen.

Show pupils a number between 20 and 50 on a 100 square. Ask pupils to show a representation of a number **more than** the number given, represented by a picture. Repeat to practise. Photograph for the next day as a reminder.

Friday

Remind pupils of Thursday's activity using the photographs. Give each pupil a whiteboard and pen.

Show pupils a number between 50 and 100 on a 100 square. Ask pupils to show a representation of a number **more than** the number given, represented by a picture. Repeat to practise.

Week 5: Representing numbers

Identify and represent numbers using number lines

Resources: place-value arrow cards, number lines from 0–100, pegs, blank number lines

Vocabulary: number, numeral, zero, one, two, three ... twenty, thirty ... one hundred, none, how many?, count, count up / on / down / back, count in ones, twos, fives and tens, many, few, odd, even, every other, how many times?, pattern, pair, ones, tens, exchange, digit, 'teen' numbers, the same number as, as many as, equal to, greater than / less than, less / more, larger, bigger, less, fewer, smaller, compare, order, size, first ... twentieth, last, before, after, next

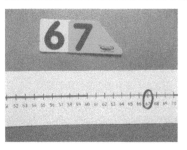

Monday

Give each pair of pupils a set of place-value arrow cards and a number line from 0–100.

Pupils take turns to choose a tens card and a ones card. They make the number and circle it on the number line.

Tuesday

Give each pair of pupils a set of place-value arrow cards, a number line from 0–100, some pegs and a whiteboard and pen.

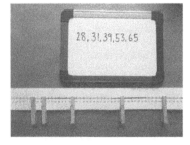

Partner 1 in each pair chooses a tens card and a ones card. They make the number and place a peg over the number on the number line to cover up the numeral. Repeat 4 times, so partner 2 has 5 numbers to work out. Partner 2 writes the concealed numbers in order from smallest to largest, using the clues on the number line and their knowledge of place value. Partner 1 checks their answers.

Wednesday

Repeat Tuesday's activity, with partners swapping roles.

Thursday

Give each pair of pupils a blank number line (this could represent 0–10, 0–20, 0–50 or 0–100).

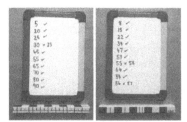

Partner 1 in each pair writes 10 numbers without partner 2 seeing and marks 10 different positions on the blank number line to represent these numbers. Partner 2 writes which numbers they think are represented on the number line. Partner 1 checks partner 2's answers.

Friday

Repeat Thursday's activity, with partners swapping roles.

Week 6: Representing numbers

Identify and represent numbers using number lines

Resources: number lines from 1–100, 100 squares, place-value arrow cards, pegs

> **Vocabulary:** number, numeral, zero, one, two, three … twenty, thirty … one hundred, none, how many?, count, count up / on / down / back, count in ones, twos, fives and tens, many, few, odd, even, every other, how many times?, pattern, pair, ones, tens, exchange, digit, 'teen' numbers, the same number as, as many as, equal to, greater than / less than, less / more, larger, bigger, less, fewer, smaller, compare, order, size, first … twentieth, last, before, after, next

Monday

Give each pair of pupils a number line from 1–100 and a 100 square.

Partner 1 in each pair circles 10 numbers on the 100 square and then reads out the numbers from smallest to largest. Partner 2 marks the numbers on the number line.

Tuesday

Repeat Monday's activity, with partners swapping roles.

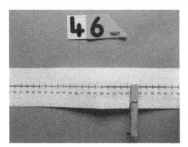

Wednesday

Give each pair of pupils a number line from 1–100, a set of place-value arrow cards and a peg.

Partner 1 in each pair chooses a tens card and a ones card and makes a two-digit number. Partner 2 places a peg over a number which is **less than** that number. Partner 1 checks that the representation is correct.

Thursday

Give each pair of pupils a number line from 1–100, a set of place-value arrow cards and a peg.

Partner 2 in each pair chooses a tens card and a ones card and makes a two-digit number. Partner 1 places a peg over a number which is **more than** that number. Partner 2 checks that the representation is correct.

Friday

Give each pupil (or pair) a number line from 1–100 and a peg.

Show the pupils a two-digit number using place-value arrow cards. Call out whether pupils should cover a number which is **less than**, **more than** or **equal to** the given number. Pupils check their answers with a partner.

Week 1: Place value

Read and write numbers to 20 in numerals

Resources: numeral cards, number lines, place-value arrow cards

Vocabulary: number, numeral, zero, one, two, three … twenty, thirty … one hundred, none, how many?, count, count up / on / down / back, count in ones, twos, fives and tens, many, few, odd, even, every other, how many times?, pattern, pair, ones, tens, digit, 'teen' numbers, the same number as, as many as, equal to, greater than / less than, less / more, larger, bigger, less, fewer, smaller, compare, order, size, first … twentieth, last, before, after, next, between, halfway between, above, below

Monday

Give each pair of pupils a set of numeral cards from 1–20.

Ask pupils to work together to place the cards in order from the smallest to the largest.

Shuffle the cards and ask the pairs to order from largest to smallest.

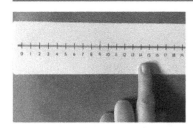

Tuesday

Give each pupil a number line from 0–20.

Ask them to count up and down the number line from 0–20.

Give a target number (e.g. 14) and count up to that number and back down to 0.

Wednesday

Give each pair of pupils a set of place-value arrow cards and a whiteboard and pen.

Call out a number between 1 and 20. Ask each pair to represent that number using place-value arrow cards and by writing the number on their whiteboard. Make sure both pupils have the opportunity to write the numerals and use the arrow cards.

Thursday

Give each pupil a set of place-value arrow cards and a whiteboard and pen.

Repeat Wednesday's activity with pupils working independently this time. Ensure lots of attention is given to the 'teen' numbers.

Friday

Give each pupil a whiteboard and pen.

Show pupils a number between 11 and 20 using a set of place-value arrow cards. Ask pupils to write the number you have shown them in the tens / ones partitioned form.

Week 2: Place value

Read and write numbers to 20 in words

Resources: number word cards, place-value arrow cards

Vocabulary: number, numeral, zero, one, two, three … twenty, thirty … one hundred, none, how many?, count, count up / on / down / back, count in ones, twos, fives and tens, many, few, odd, even, every other, how many times?, pattern, pair, ones, tens, digit, 'teen' numbers, the same number as, as many as, equal to, greater than / less than, less / more, larger, bigger, less, fewer, smaller, compare, order, size, first … twentieth, last, before, after, next, between, halfway between, above, below

one
two
three
four

Monday

Give each pair of pupils a set of number word cards from 1–20.

Ask each pair to place the word cards in order from the smallest to the largest.

fourteen
thirteen
twelve
eleven

Tuesday

Give each a pair of pupils a set of number word cards from 1–20.

Ask each pair to place the word cards in order from the largest to the smallest.

Wednesday

Give each pair of pupils a set of place-value arrow cards and a whiteboard and pen.

Call out a number between 1 and 20. Ask each pair to represent that number using place-value arrow cards and by writing the number on their whiteboard in words. Ensure both pupils have the opportunity to write the number in words and to use the cards.

Thursday

Give each pupil a set of place-value arrow cards and a whiteboard and pen.

Repeat Wednesday's activity with pupils working independently this time. Ensure lots of attention is given to the 'teen' numbers.

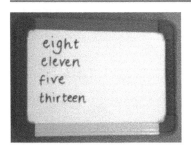

Friday

Give each pupil a whiteboard and pen.

Show pupils a number between 1 and 20 using a set of place-value arrow cards. Ask pupils to write the number in word form. Repeat as many times as possible in the time available.

Week 3: Place value

Read and write numbers to 50 in numerals

Resources: numeral cards, place-value arrow cards, objects, containers, a tin

Vocabulary: number, numeral, zero, one, two, three … twenty, thirty … one hundred, none, how many?, count, count up / on / down / back, count in ones, twos, fives and tens, many, few, odd, even, every other, how many times?, pattern, pair, ones, tens, digit, 'teen' numbers, the same number as, as many as, equal to, greater than / less than, less / more, larger, bigger, less, fewer, smaller, compare, order, size, first … twentieth, last, before, after, next, between, halfway between, above, below

Monday

Give each pair of pupils a set of numeral cards from 1–50 and a whiteboard and pen.

Pupils take turns to pick a numeral card from the pile and read it aloud. Without looking at the card, their partner must write the number on the whiteboard. Together they check the answer is correct and swap over. (Pupils could use Dienes to represent the number too.)

Tuesday

Give each pair of pupils a set of numeral cards from 1–50 and a set of place-value arrow cards.

Pupils take turns to pick a numeral card from the pile and read it aloud. Without looking at the card, their partner must make the number using the place-value arrow cards. Together they check the answer is correct and swap over.

Wednesday

Give each pair of pupils a set of objects, a container and a whiteboard and pen.

Call out a number between 1 and 50. Each pair counts out that number of objects. One pupil writes the number in numerals while the other pupil writes the number in its tens / ones partitioned form. Repeat, with pupils swapping roles.

Thursday

Give each pupil a whiteboard and pen.

Drop objects into a tin one-by-one, while everyone counts together. When you stop dropping objects into the tin, pupils write that number on their whiteboards in numerals.

Friday

Give each pupil a whiteboard and pen.

Begin by showing that you have 50 objects in a tin. Then, counting together, count back one-by-one. When you stop removing objects from the tin, pupils write that number on their whiteboards in numerals.

Week 4: Place value

Count, read and write numbers to 100 in numerals

Resources: Dienes sets, numeral cards, place-value arrow cards

Vocabulary: number, numeral, zero, one, two, three … twenty, thirty … one hundred, none, how many?, count, count up / on / down / back, count in ones, twos, fives and tens, many, few, odd, even, every other, how many times?, pattern, pair, ones, tens, digit, 'teen' numbers, the same number as, as many as, equal to, greater than / less than, less / more, larger, bigger, fewer, smaller, compare, order, size, first … twentieth, last, before, after, next, between, halfway between, above, below

Monday

Give each pair of pupils a set of Dienes.

Show pupils a numeral card between 1 and 100. Each pair reads the number and uses Dienes to represent that number.

Repeat as many times as possible in the time available.

Tuesday

Give each pair of pupils a set of Dienes.

Show pupils a number between 1 and 100 using place-value arrow cards. Each pair reads the number and uses Dienes to represent that number.

Repeat as many times as possible in the time available.

Wednesday

Give each pair of pupils a set of place-value arrow cards.

Show pupils a number between 1 and 100 using Dienes. Each pair reads the number and uses the place-value arrow cards to represent that number.

Repeat as many times as possible in the time available.

Thursday

Give each pair of pupils a set of numeral cards from 1–100 and a whiteboard and pen.

Pupils take turns to pick a numeral card from the pile and read it aloud. Without seeing the card, their partner must write the number on the whiteboard. Together they check the answer is correct and swap over.

Friday

Give each pair of pupils a set of numeral cards from 1–100 and a set of place-value arrow cards.

Pupils take turns to pick a numeral card from the pile and read it aloud. Without seeing the card, their partner must make the number using the place-value arrow cards. Together they check the answer is correct and swap over.

Week 5: Place value

Given a number, identify one more

Resources: objects, place-value arrow cards, 100 squares

Vocabulary: number, numeral, zero, one, two, three … twenty, thirty … one hundred, none, how many?, count, count up / on / down / back, count in ones, twos, fives and tens, many, few, odd, even, every other, how many times?, pattern, pair, ones, tens, digit, 'teen' numbers, the same number as, as many as, equal to, greater than / less than, less / more, larger, bigger, less, fewer, smaller, compare, order, size, first … twentieth, last, before, after, next, between, halfway between, above, below

Monday

Give each pair of pupils a set of everyday objects.

Use place-value arrow cards to show pupils a number between 1 and 50. Pupils count out this number of objects and then add one more to find the number that is one more.

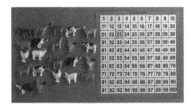

Tuesday

Give each pair of pupils a 100 square.

Count a set of objects together. When you stop counting, pupils circle the number on their 100 square that is one more than the number you counted to.

Wednesday

Give each pupil a whiteboard and pen.

Count a set of objects together. When you stop counting, pupils write in numerals the number that is one more than the number you counted to.

Thursday

Give each pair of pupils a set of place-value arrow cards.

Count a set of objects together. When you stop counting, pupils use the place-value arrow cards to show one more than the number you counted to.

Friday

Give each pupil a set of place-value arrow cards.

Repeat Thursday's activity, with pupils working independently this time.

Week 6: Place value

Given a number, identify one less

Resources: objects, place-value arrow cards, 100 squares, containers

Vocabulary: number, numeral, zero, one, two, three … twenty, thirty … one hundred, none, how many?, count, count up / on / down / back, count in ones, twos, fives and tens, many, few, odd, even, every other, how many times?, pattern, pair, ones, tens, digit, 'teen' numbers, the same number as, as many as, equal to, greater than / less than, less / more, larger, bigger, less, fewer, smaller, compare, order, size, first … twentieth, last, before, after, next, between, halfway between, above, below

---

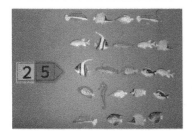

Monday

Give each pair of pupils a set of everyday objects.

Use place-value arrow cards to show pupils a number between 1 and 50. Pupils count out the number of objects shown on the cards. They then remove one object to find one fewer.

Tuesday

Give each pair of pupils a 100 square and a set of everyday objects.

Pupils count out a given number of objects into a container. Count backwards together from the given number, removing one object each time. When you stop counting, pupils circle the number on their 100 square that is one less than the number you counted backwards to.

Wednesday

Give each pupil a whiteboard and pen.

Place a given number of objects in a container. Count backwards together from the given number, removing one object each time. When you stop counting, pupils write in numerals the number that is one less than the number you counted backwards to.

Thursday

Give each pair of pupils a set of place-value arrow cards. Display a large 100 square.

Place a given number of objects in a container. Count backwards together from the given number, removing one object each time. When you stop counting, pupils use the place-value arrow cards to show one less than the number you counted backwards to.

Friday

Give each pupil a set of place-value arrow cards. Display a large 100 square.

Repeat Thursday's activity with pupils working independently this time.

Read, write and interpret mathematical statements involving + and = signs

Resources: dominoes, large sheets of paper with squares labelled 1–12

Vocabulary: +, add, addition, more, more than, plus, make, sum, total, altogether, how many more to make …?, how many more is … than …?, how much more is …?, −, subtract, take / take away, minus, less, one less, two less, ten less …, how many fewer is … than …?, how much less is … than …?, what is the difference between …?, =, equals / equal to, is the same as, sign, symbol

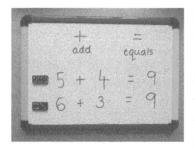

Monday

Draw the + symbol on the board and remind pupils that this means we add, or combine, two amounts. Draw the = symbol on the board and explain that it means 'equals' or the 'same as'.

Take one domino from your set. Count the number of spots on each side of the domino and write the number sentence for that domino. Count the total number of spots. Complete several together as a group.

Tuesday

Give each pair of pupils a handful of dominoes and a whiteboard and pen.

Repeat Monday's activity, with partner 1 in each pair writing the number sentence and partner 2 finding the answer by counting the spots.

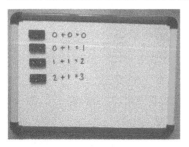

Wednesday

Give each pair of pupils a handful of dominoes and a whiteboard and pen.

Repeat Tuesday's activity, with pupils swapping roles.

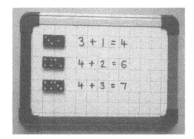

Thursday

Give each pair of pupils a set of dominoes and a whiteboard and pen.

Challenge pupils to find dominoes with a total number of spots from 0–20. Can they go all the way to 20 using just one set of dominoes? If not, what is the largest number they can get to? (12)

Pupils should write the totals as a number sentence; encourage them to use a systematic way of working.

Friday

Give each pair of pupils a set of dominoes, a large sheet of paper with large squares labelled 1–12 (or use cups, bun cases, whiteboards or chalk in the playground) and a whiteboard and pen.

Give each pair 10 minutes to correctly place all 28 dominoes in the correct spot total.

Week 2: Addition and subtraction

Read, write and interpret mathematical statements involving – and = signs

Resources: dominoes

Vocabulary: +, add, addition, more, more than, plus, make, sum, total, altogether, how many more to make …?, how many more is … than …?, how much more is …?, –, subtract, take / take away, minus, less, one less, two less, ten less …, how many fewer is … than …?, how much less is … than …?, what is the difference between …?, =, equals / equal to, is the same as, sign, symbol

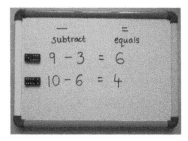

Monday

Draw the – symbol on the board and remind pupils that this means we subtract, or take away, from an amount. Draw the = symbol on the board and remind pupils that it means 'equals' or the 'same as'.

Take one domino from your set. Count the total number of spots on the domino. Explain that you are going to take away the spots on one side of the domino from the spot total. Now count the number of spots remaining. Write the number sentence for that domino. Complete several together as a group.

Tuesday

Give each pair of pupils a handful of dominoes and a whiteboard and pen.

Repeat Monday's activity, with partner 1 in each pair writing the number sentence and partner 2 finding the answer by counting the spots.

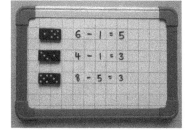

Wednesday

Give each pair of pupils a handful of dominoes and a whiteboard and pen.

Repeat Tuesday's activity, with pupils swapping roles.

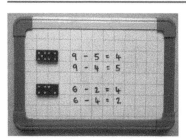

Thursday

Give each pair of pupils a handful of dominoes and a whiteboard and pen.

Ask each pupil in the pair to take a domino. Explain that they are going to write two subtraction facts for their domino. Model this and then allow pupils time to practise. Pupils check each other's work.

Friday

Give each pair of pupils a handful of dominoes and a whiteboard and pen.

Ask each pupil in the pair to take a domino. Explain that they are going to write two subtraction facts and two addition facts for their domino. Model this and then allow pupils time to practise. Pupils check each other's work.

Week 3: Addition and subtraction

Represent and use number bonds and related subtraction facts within 20 (to 10)

Resources: cubes

Vocabulary: +, add, addition, more, more than, plus, make, sum, total, altogether, how many more to make …?, how many more is … than …?, how much more is …?, −, subtract, take / take away, minus, less, one less, two less, ten less …, how many fewer is … than …?, how much less is … than …?, what is the difference between …?, =, equals / equal to, is the same as, sign, symbol

Monday

Give each pair of pupils a set of 10 cubes.

Model how to make all the addition and subtraction number bonds within 10. Also introduce '10 + 0'. Allow pupils time to explore the number bonds in pairs.

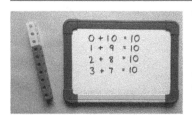

Tuesday

Give each pupil a tower of 10 cubes and a whiteboard and pen.

Ask them to write number sentences to show all of the addition number bonds within 10. Encourage a systematic way of recording.

Wednesday

Give each pupil a tower of 10 cubes.

Model removing 1 cube to leave 9 cubes. Model saying the sentence to make clear the vocabulary of subtraction.

Ask pupils to do the same, starting with 10 cubes and removing 1 cube, then 2 cubes, etc. Encourage the use of the word 'zero' or 'none' to show that 10 cubes subtract 0 cubes is 10 cubes.

Thursday

Give each pupil a tower of 10 cubes and a whiteboard and pen.

Pupils work in pairs. Partner 1 chooses a number between 5 and 9. Both pupils use their cubes to make a tower representing that number. They find as many addition and subtraction facts as they can by separating the cubes and recording as on Wednesday.

Friday

Repeat Thursday's activity, with pupils swapping roles. Highlight that there are lots of different ways in which each number can be represented, using addition and subtraction facts.

Represent and use number bonds and related subtraction facts within 20

Resources: cubes, 20-bead strings

> **Vocabulary:** +, add, addition, more, more than, plus, make, sum, total, altogether, how many more to make …?, how many more is … than …?, how much more is …?, −, subtract, take / take away, minus, less, one less, two less, ten less …, how many fewer is … than …?, how much less is … than …?, what is the difference between …?, =, equals / equal to, is the same as, sign, symbol

Monday

Give each pair of pupils a set of 20 cubes.

Ask them to make towers to represent all of the numbers from 1–9. Once they have done this, they should make a tower of 10 cubes and add 1, 2, 3, etc. to make the numbers from 11–20.

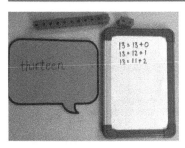

Tuesday

Give each pair of pupils 20 cubes as two towers of 10 (in two different colours) and a whiteboard and pen.

Call out a number between 11 and 20. Pupils work in pairs to create that number using the cubes. They then write as many addition and subtraction number sentences as they can for the number, through manipulating the set of cubes.

Wednesday

Give each pupil a whiteboard and pen.

Show a 20-bead string. Count the beads, from 0–20. Move half the beads along the string and show that 10 + 10 = 20 and 20 − 10 = 10.

Use the beads to show addition facts from 20 + 0 to 0 + 20. As you model this, pupils write the number sentences in the format 20 = 20 + 0, 20 = 19 + 1, 20 = 18 + 2, etc. Repeat for subtraction facts (e.g. 20 − 1 = 19).

Thursday

Give each pair of pupils a 20-bead string and a whiteboard and pen.

Call out a number between 1 and 20. Partner 1 in each pair shows that number of beads and partner 2 says how many more beads will give the total 20. They prove the fact to their partner by counting the beads. Pupils then give two subtraction facts (e.g. 'twenty subtract nine equals eleven', and 'twenty subtract eleven equals nine').

Friday

Repeat Thursday's activity, with pupils swapping roles.

Pupils could record their number sentences on a whiteboard.

Week 5: Addition and subtraction

Add one-digit numbers to 20

Resources: objects, containers, 1–6 dice

Vocabulary: +, add, addition, more, more than, plus, make, sum, total, altogether, how many more to make …?, how many more is … than …?, how much more is …?, −, subtract, take / take away, minus, less, one less, two less, ten less …, how many fewer is … than …?, how much less is … than …?, what is the difference between …?, =, equals / equal to, is the same as, sign, symbol

Monday

Set out bowls each containing between 1 and 9 objects.

Pupils collect two bowls and find the total number represented by the two sets of objects. To do this, they combine the two amounts and count the whole set to find the total.

Tuesday

Set out bowls each containing between 1 and 9 objects. Give each pupil a whiteboard and pen.

Pupils collect two sets of objects and add the two totals together using the vocabulary of 'add' and 'plus'. They record their number sentence on a whiteboard.

Wednesday

Set out bowls each containing between 1 and 9 objects.

Pupils collect two sets of objects. This time they find the total by counting on from the first set of objects.

Thursday

Set out bowls each containing between 1 and 9 objects. Give each pupil a whiteboard and pen.

Pupils collect two sets of objects and add the two totals together, using the vocabulary of 'more than'. They record their number sentence on a whiteboard.

Friday

Set out bowls each containing between 1 and 9 objects. You should also have some 'extra' objects so that pupils can use these if they are not ready to use their fingers to count on.

Pupils collect a set of objects and a 1–6 dice. They count the number of objects in the bowl, then roll the dice and 'count on' that amount to find the total.

Week 6: Addition and subtraction

Subtract one-digit numbers to 20

Resources: objects, containers, 1–6 dice

Vocabulary: +, add, addition, more, more than, plus, make, sum, total, altogether, how many more to make …?, how many more is … than …?, how much more is …?, –, subtract, take / take away, minus, less, one less, two less, ten less …, how many fewer is … than …?, how much less is … than …?, what is the difference between …?, =, equals / equal to, is the same as, sign, symbol

Monday

Set out bowls each containing between 8 and 10 objects.

Pupils collect a set of objects and a 1–6 dice. They count the number of objects and roll the dice to find out how many they should remove from the bowl. They then count the amount left in the bowl. They are subtracting by taking away (removing).

Tuesday

Set out bowls each containing between 12 and 20 objects and 1–6 dice.

Pupils collect a set of objects and a dice. They count the number of objects and roll the dice to find out how many they should remove from the bowl. They then count the amount left in the bowl. They are subtracting by taking away (removing).

Wednesday

Give each pupil or pair a whiteboard and pen.

Repeat Tuesday's activity, in pairs or individually (depending on pupils' confidence). Pupils record what they do as number sentences on a whiteboard.

Thursday

Have available a range of different objects and containers.

Ask pairs of pupils to choose 8 objects of one type and 3 objects of another type.

Demonstrate how to place the objects side by side and compare them to find the difference. Model writing the sentence. Allow time for the pairs to practise this.

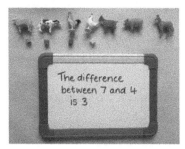

Friday

Give each pupil or pair a whiteboard and pen.

Repeat Thursday's activity, in pairs or individually (depending on pupils' confidence). Pupils record what they do as number sentences on a whiteboard.

Week 1: Multiplication and division

Resources: objects, elastic bands, cubes, bowls, large-squared paper

Vocabulary: multiplying, lots of, grouping, groups of, times, multiply, multiplied by, repeated addition, array, row, column, double, halve, share, share equally, one each, two each, three each …, group in pairs, twos, fives, tens, equal groups of, divide, divided by, divided into, left, left over

Monday

Give each pair of pupils a set of 10 pairs of objects (e.g. animals) banded together in pairs.

Show 2 animals banded together. Explain that two can also be called a pair. Tell pupils you are going to count in twos to find out how many objects there are altogether. Model counting them one-by-one afterwards. Ask pupils to repeat with different quantities of banded objects.

Tuesday

Give each pair of pupils 5 pairs of banded objects and a whiteboard and pen.

Ask them to count in twos to find out how many objects there are altogether. Ask pupils to write the total. (Allow them to do this independently to see how they complete the activity; they will use a variety of ways.)

Wednesday

Explain that you are going to count objects in twos, using an array to record the totals. Model a 2 × 5 array. Explain that an array must have the same number in each row / column and be spaced evenly.

Give each pair of pupils 10 cubes in a bowl. Ask pupils to create their own array by putting the cubes in rows of two on large-squared paper (or a whiteboard). Ask them to tell each other about the array.

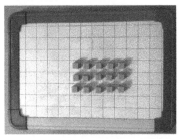

Thursday

Count in fives from 0–15, modelling a 5 × 3 array to record the groups of 5.

Give each pair of pupils 15 cubes in a bowl. Ask pupils to create their own array by putting the cubes in rows on large-squared paper (or a whiteboard). Ask them to tell each other about the array.

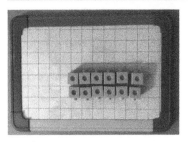

Friday

Allow pupils to choose their own objects from a selection available in the classroom.

They should choose a multiple of two or five (or give each pupil a multiple) and create their own array on large-squared paper (or a whiteboard).

Pupils pair with another pupil and tell each other about their arrays.

Week 2: Multiplication and division

Resources: pencils, pots, objects, large pieces of paper

Vocabulary: multiplying, lots of, grouping, groups of, times, multiply, multiplied by, repeated addition, array, row, column, double, halve, share, share equally, one each, two each, three each …, group in pairs, twos, fives, tens, equal groups of, divide, divided by, divided into, left, left over

Monday

Prepare pots of 8 pencils. Pupils will work in threes.

Take 8 pencils and show pupils how to share them equally between two pupils: ask two pupils to stand at the front and share 'one for you, one for you, one for you, one for you', etc. Ask how many pencils each pupil has got.

Split pupils into groups of three: one pupil shares out the pencils and the other two receive and count the pencils.

Tuesday

Prepare pots of 12 pencils.

Model the sharing process again. Ask pupils to work in threes again: one pupil shares out the pencils and the other two receive and count the pencils. Start with 10 pencils, then repeat with 6 pencils and 12 pencils.

Wednesday

Prepare pots of 12 pencils. Give each pupil a whiteboard and draw a line down the middle so they can 'share' between two.

Start with pupils selecting 4 pencils and sharing them using their whiteboard to see how many are in each group. Repeat with 8, 10 and 12 pencils.

Thursday

Prepare pots of 10 pencils.

Explain that today you are going to divide by grouping. Take a pot of 10 pencils and model grouping the pencils into twos.

Give each pair of pupils a pot of 10 pencils. Ask them to do the same. Repeat by grouping into fives.

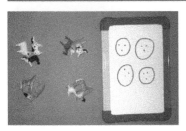

Friday

Choose an everyday object that pupils are familiar with. Give pairs of pupils a set of 12 objects, a large piece of paper and a whiteboard and pen.

Pupils work in pairs to group the objects in twos, threes, fours and sixes, placing them on the paper and drawing around the groups each time. One pupil in each pair should draw a pictorial representation, alongside the concrete representation.

Week 3: Multiplication and division

Make connections between arrays, number patterns and counting in twos

Resources: 100 squares, cubes, bowls

> **Vocabulary:** multiplying, lots of, grouping, groups of, times, multiply, multiplied by, repeated addition, array, row, column, double, halve, share, share equally, one each, two each, three each ..., group in pairs, twos, fives, tens, equal groups of, divide, divided by, divided into, left, left over

Monday

Give each pupil a 100 square.

Model how to start at 0 and count in twos up to 30, colouring in or marking each multiple of two. Ask pupils if they can spot a pattern.

Ask pupils if they can think of things that come in twos (pairs), noting ideas (e.g. twins, socks, eyes).

Tuesday

Give each pupil a blank 100 square.

Together, count in twos to 50 and back again, with pupils circling the numbers as they count along.

Discuss patterns that they can see. Are the patterns the same as on Monday?

Wednesday

Give each pupil a blank 100 square.

Together, count in twos to 100 and back again, with pupils circling the numbers as they count along.

Discuss patterns that they can see. Explain that we call the multiples of two 'even' numbers.

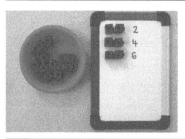

Thursday

Give each pupil a bowl of 20 cubes and a whiteboard and pen.

Ask pupils to put the cubes into twos, setting them out in pairs on their whiteboard. They should then write their 'counting in twos' numbers alongside.

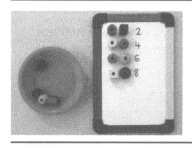

Friday

Place different **even** quantities of cubes in bowls.

Ask pupils to count the quantities in twos, writing the totals on their whiteboards.

Make connections between arrays, number patterns and counting in fives

Resources: 100 squares, cubes, bowls

Vocabulary: multiplying, lots of, grouping, groups of, times, multiply, multiplied by, repeated addition, array, row, column, double, halve, share, share equally, one each, two each, three each …, group in pairs, twos, fives, tens, equal groups of, divide, divided by, divided into, left, left over

Monday

Give each pupil a 100 square.

Model how to start at 0 and count in fives up to 30, colouring in or marking each multiple of five.

Ask pupils if they can spot a pattern. Ask pupils if they can think of things that come in fives, noting ideas (e.g. fingers, toes, 5p coins, £5 notes, 5 minutes on a clock).

Tuesday

Give each pupil a blank 100 square. Together, count to 50 and back again in fives, with pupils circling the numbers as they count along.

Discuss patterns that they can see. Are the patterns the same as on Monday?

Wednesday

Give each pupil a blank 100 square. Together, count to 100 and back again in fives, with pupils circling the numbers as they count along.

Discuss patterns that they can see. Are the patterns the same as on previous days? Can they find a rule for when we count in fives? (The number always has a 0 or 5 ones.)

Thursday

Give each pupil a bowl of 30 cubes and a whiteboard and pen.

Ask pupils to put the cubes into fives, setting them out in fives on their whiteboard. They then write their 'counting in fives' numbers alongside.

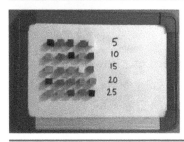

Friday

Place different quantities of cubes in bowls, in multiples of five. Give each pupil a bowl of cubes and a whiteboard and pen.

Ask pupils to create arrays where the rows are in fives. They should write their 'counting in fives' numbers alongside to find the total.

Week 5: Multiplication and division

Make connections between arrays, number patterns and counting in tens

Resources: 100 squares, Dienes sets, 10p coins, bowls

Vocabulary: multiplying, lots of, grouping, groups of, times, multiply, multiplied by, repeated addition, array, row, column, double, halve, share, share equally, one each, two each, three each …, group in pairs, twos, fives, tens, equal groups of, divide, divided by, divided into, left, left over

Monday

Give each pupil a 100 square.

Model how to start at 0 and count in tens up to 100, colouring in or marking each multiple of ten. Ask pupils if they can spot a pattern.

Ask pupils if they can think of things that come in tens, noting ideas (e.g. fingers, toes, 10p coins, £10 notes, 10 legs on a crab, 10 bowling pins).

Tuesday

Give each pair of pupils a set of 10 rods and a 100 block from a set of Dienes.

Together, count in tens from 0–100, placing the rods on top of the block so pupils can see the numbers getting closer to 100.

Count back again, removing a rod of 10 each time so that pupils can see the numbers getting closer to 0. Practise several times, counting to different target numbers.

Wednesday

Give each pair of pupils 10 × 10p coins in a bowl.

Together, count in multiples of 10p up to 100p; tell pupils that this is also called £1.

Practise counting to different target amounts (e.g. 80p, 60p, 90p) and backwards to 0.

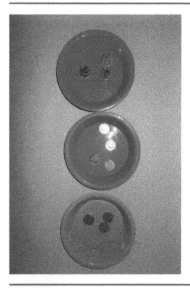

Thursday

Place different amounts of 10p coins in bowls, enough for one bowl for each pair of pupils. Give each pair of pupils a whiteboard and pen.

Seat pupils in a large circle. Give each pair a bowl and ask them to place the coins on their whiteboards and to write the 'counting in tens' numbers alongside. Once everyone has completed their bowl, rotate the bowls around the circle.

Friday

Place different amounts of 10p coins in bowls, enough for one bowl for each pupil. Give each pupil a whiteboard and pen.

Repeat Thursday's activity, but this time pupils should work independently.

Week 6: Multiplication and division

Resources: cubes, 10p coins, 20-bead strings

Vocabulary: multiplying, lots of, grouping, groups of, times, multiply, multiplied by, repeated addition, array, row, column, double, halve, share, share equally, one each, two each, three each …, group in pairs, twos, fives, tens, equal groups of, divide, divided by, divided into, left, left over

Monday

Explain that doubling is when you have two lots of the same quantity (multiplying by two).

Give each pupil a set of 5 cubes. Give each pair of pupils a whiteboard and pen.

Ask pupils to draw a line down the middle of their whiteboard. Ask partner 1 in each pair to place 3 cubes on one side of the line, and then ask partner 2 to place 3 cubes on the other side of the line. Explain that they have doubled 3. Ask what the total is. Repeat for numbers from 1–5.

Tuesday

Give each pupil a set of 10 cubes. Give each pair of pupils a whiteboard and pen.

Ask pupils to draw a line down the middle of their whiteboard. Ask partner 1 in each pair to place 6 cubes on one side of the line, and then ask partner 2 to place 6 cubes on the other side of the line. Explain that they have doubled 6. Ask what the total is. Repeat for numbers from 1–10.

Wednesday

Give each pair of pupils 10 × 10p coins and a whiteboard and pen. Ask pupils to take 5 × 10p coins each.

Ask pupils to draw a line down the middle of their whiteboard. Ask partner 1 in each pair to place 2 × 10p coins on one side of the line, and then ask partner 2 to place 2 × 10p coins on the other side of the line. Explain that they have doubled 20p. Ask what the total is. Repeat for 10p, 30p, 40p and 50p.

Thursday

Give each pupil a 20-bead string.

Ask pupils to slide 5 beads to one side. Ask them to double the number of beads by sliding an additional 5 beads across. Count the total. Repeat to give practice of doubling other numbers from 1–10.

Friday

Give each pupil a whiteboard and pen.

Ask pupils to draw a line down the middle of their whiteboard. Call out a number between 1 and 10 and ask pupils to draw dots to represent the quantity on one side of the line, and then to draw the same amount of dots on the other side of the line to calculate and represent the double.

Week 1: Fractions

Recognise, find and name a half of a shape

Resources: an apple, paper shapes, 2-D plastic shapes, whiteboard pens

Vocabulary: halve, share, share equally, one each, two each, three each ..., group in pairs, twos, fours ..., equal groups of, divide, divided by, divided into, left, left over, part, equal parts, fraction, one, one whole, one half, two halves, one quarter ... four quarters

Monday

Explain that halving is when you share, equally, into two parts.

Show an apple and ask two pupils who like apples to come out. Say you are going to cut the apple in half and give half to each pupil. Cut the apple unequally. Offer a piece to each pupil and ask what they think about the halves.

Give each pupil a paper square. See how many ways they can halve it by folding.

Tuesday

Give each pupil a rectangular paper shape.

See how many ways they can halve it. Share the different ways with the group.

Wednesday

Give each pair of pupils a paper plate (or a circle of paper) and ask them to explore the different ways in which their circle can be halved.

Ask questions about the shapes (e.g. Do all of the halves look the same? When you compare halves with someone else, are they the same?).

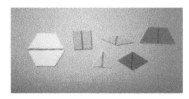

Thursday

Give each pair of pupils a selection of 2-D plastic shapes and a whiteboard pen.

Ask them to draw a line that will halve each shape.

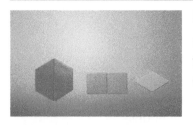

Friday

Give each pair of pupils sets of two identical 2-D plastic shapes and explain that they have been given two halves. Ask them to put the two shapes together to create one whole shape.

Recognise, find and name a half of a set of objects

Resources: objects, bowls, paper

Vocabulary: halve, share, share equally, one each, two each, three each …, group in pairs, twos, fours …, equal groups of, divide, divided by, divided into, left, left over, part, equal parts, fraction, one, one whole, one half, two halves, one quarter … four quarters

Monday

Remind pupils that halving is when you equally share, or divide, a group of objects into two equal parts.

Give each pair of pupils a set of 10 objects and a whiteboard and pen.

Ask pupils to draw a line down the middle of their whiteboard and to share the 10 objects between the two sections. Explain that they have halved the objects. Ask how many are in each section. Repeat with 8, 6, 4 and 2 objects.

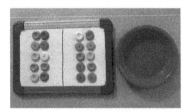

Tuesday

Give each pair of pupils a set of 20 objects and a whiteboard and pen.

Ask each pair to draw a line down the middle of their whiteboard and to share the 20 objects between the two sections. Explain that they have halved the objects. Ask how many are in each section. Repeat with 18, 16, 14 and 12 objects.

Wednesday

Give each pair of pupils a whiteboard and a sheet of paper. Set out bowls containing different (even) sets of objects around the classroom.

Explain that there are different whole amounts of objects in each bowl. Ask pupils to work in pairs to find half of the different amounts by sharing equally and to record their answers.

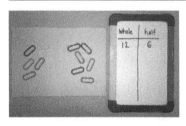

Thursday

Give each pair of pupils a sheet of paper and a whiteboard and pen.

Ask them to fold the paper in half. Set out different (even) sets of objects around the classroom. Explain that there are different whole amounts of objects in each bowl. Ask pupils to work together to find half of the different amounts, and record their answers.

Friday

Give each pair of pupils a set of objects (between 1 and 10 objects), a sheet of paper and a whiteboard and pen.

Explain that they have been given half of an amount of objects. Ask them to place the objects on one half of their paper. How many would be in a whole set? They write the whole and half numbers on their whiteboard. Repeat with different sets.

Week 3: Fractions

Recognise, find and name a half of a quantity

Resources: pegboards, pegs, paper, 1p coins, bowls

Vocabulary: halve, share, share equally, one each, two each, three each ..., group in pairs, twos, fours ..., equal groups of, divide, divided by, divided into, left, left over, part, equal parts, fraction, one, one whole, one half, two halves, one quarter ... four quarters

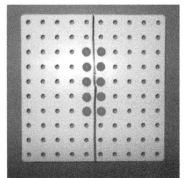

Monday

Remind pupils that halving is when you equally share something into two parts. This week will focus on halving numbers.

Give each pair of pupils a pegboard with a line drawn down the middle and 10 pegs.

Model how to share the pegs between the two sections. Ask the pairs to halve their 10 pegs. Ask how many are in each section. Repeat with 12 pegs, 14 pegs, etc.

Tuesday

Give each pair of pupils a pegboard with a line down the middle. Set out different quantities of pegs for pupils to investigate.

Repeat Monday's activity to practise halving.

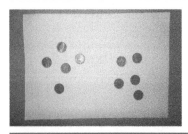

Wednesday

Give each pair of pupils a sheet of A4 paper and 10p in 1p coins.

Ask each pair to fold their paper in half and to share the 1p coins equally between the two sections. Ask how many coins are in each half. Summarise that one half of 10p is 5p.

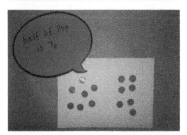

Thursday

Put different even amounts of 1p coins into bowls for pupils to investigate.

Give each pair of pupils a sheet of paper and ask them to fold it in half.

Ask the pairs to find half of the 1p coins in the same way that they did on Wednesday. Once they have found half, they should tell each other 'half of ___p is ___p'.

Friday

Give each pair of pupils a sheet of paper and a whiteboard and pen.

They repeat Thursday's activity, recording the whole amount and half of the whole amount. They check the answers together.

Week 4: Fractions

Recognise, find and name a quarter of a shape

Resources: an orange, paper shapes, 2-D plastic shapes, bowls

> **Vocabulary:** halve, share, share equally, one each, two each, three each …, group in pairs, twos, fours …, equal groups of, divide, divided by, divided into, left, left over, part, equal parts, fraction, one, one whole, one half, two halves, one quarter … four quarters

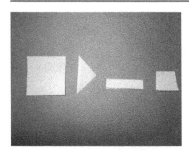

Monday

Explain that a quarter is when you share, equally, into four parts.

Show an orange and ask four pupils who like oranges to come out to the front. Say you are going to cut the orange into quarters and give one quarter to each of them. Cut the orange unequally. Offer a piece to each pupil and ask what they think about the quarters.

Give each pupil a paper square. See how many ways they can fold it into quarters.

Tuesday

Give each pupil a paper rectangle.

See how many ways they can fold it into quarters – you could model halving, then halving again to find a quarter. Share the different ways with the group.

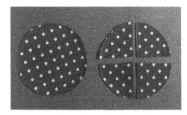

Wednesday

Give each pair of pupils a paper plate (or a circle of paper) and ask them to explore the different ways in which their circle can be quartered.

Ask questions about the shapes, e.g. Do all of the quarters look the same? When you compare quarters with someone else, are they the same?

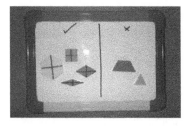

Thursday

Give each pair of pupils a selection of 2-D plastic shapes and a whiteboard and pen.

Ask them to draw two lines that will quarter the shape. Afterwards, pupils should sort the shapes into those that can be quartered and those that cannot.

Friday

Give each pair of pupils four identical 2-D plastic shapes.

Ask them to put the four shapes together to create one whole new shape.

Week 5: Fractions

Recognise, find and name a quarter of a set of objects

Resources: objects, bowls, paper

Vocabulary: halve, share, share equally, one each, two each, three each …, group in pairs, twos, fours …, equal groups of, divide, divided by, divided into, left, left over, part, equal parts, fraction, one, one whole, one half, two halves, one quarter … four quarters

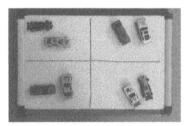

Monday

Remind pupils that a quarter is when you equally share a group of objects into four parts.

Give each pair of pupils a set of 8 objects and a whiteboard and pen.

Ask pupils to split their whiteboard into four and to share the objects between the four sections. Explain that they have quartered the objects. Ask how many are in each section. Repeat with 12 objects, 16 objects and 20 objects.

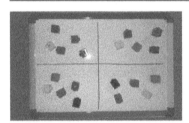

Tuesday

Give each pair of pupils a set of 20 different objects and a whiteboard and pen.

Ask pupils to split their whiteboard into four and to share the objects between the four sections. Explain that they have quartered the objects. Ask how many are in each section. Repeat with 16 objects, 12 objects and 8 objects.

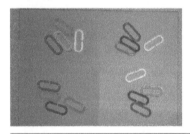

Wednesday

Give each pair of pupils a sheet of paper and ask them to fold it into quarters. Set out different sets of objects around the room, in multiples of four.

Explain that there are different whole amounts of objects in each bowl. Pupils work in pairs to find a quarter of the different amounts.

Thursday

Remind pupils that, when we find one quarter, we share, or divide, an amount into four equal parts.

Give each pupil a sheet of paper and ask them to fold it into quarters. Set out different sets of objects around the room, in multiples of four.

Explain that there are different whole amounts of objects in each bowl. Pupils work individually to find one quarter of the different amounts. Pupils could record the whole amount and quarter on a whiteboard.

Friday

Give each pair of pupils a set of objects (between 1 and 5 objects) and a whiteboard and pen.

Explain that they have one quarter of an amount of objects. Ask them to place the objects on one quarter of their whiteboard. How many would be in a whole set? Encourage pupils to represent the remaining three quarters using dots.

Week 6: Fractions

Recognise, find and name a quarter of a quantity

Resources: pegboards, pegs, paper, 1p coins, bowls

Vocabulary: halve, share, share equally, one each, two each, three each …, group in pairs, twos, fours …, equal groups of, divide, divided by, divided into, left, left over, part, equal parts, fraction, one, one whole, one half, two halves, one quarter … four quarters

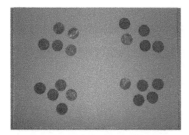

Monday

Remind pupils that a quarter is when you equally share something into four parts. This week will focus on finding quarters of a quantity.

Give each pair of pupils a pegboard with a line drawn down the middle and across and 12 pegs.

Model how to share the pegs between the four sections. Ask each pair to quarter their 12 pegs. Ask how many are in each section. Repeat with 16 pegs.

Tuesday

Give each pair of pupils a pegboard with a line drawn down the middle and across. Set out different quantities of pegs (in multiples of four) for pupils to investigate.

Repeat Monday's activity to practise finding one quarter of a quantity.

Wednesday

Give each pair of pupils a sheet of A4 paper and 20p in 1p coins.

Ask each pair to fold their paper into quarters.

Ask them to share the 1p coins equally between the four sections. Ask how many coins are in each quarter. Summarise that one quarter of 20p is 5p.

Thursday

Put different even amounts of 1p coins into bowls, in multiples of four, for pupils to investigate. Give each pair of pupils a sheet of paper.

Ask pupils to fold their paper into quarters.

Ask pairs to find one quarter of the different totals in the same way that they did on Wednesday. Once they have found one quarter they should tell each other 'one quarter of ___p is ___p'.

Friday

Give each pair of pupils a sheet of paper and a whiteboard and pen.

They repeat Thursday's activity, recording on whiteboards the whole amount and one quarter of the whole amount. They check the answers together.

Blank vocabulary cards